I AM EARTH
I AM YOUR HOME
I AM FRAGILE
TREAT ME WITH CARE

The awareness of the problems caused by humans, and practical solutions to care for distressed planet, Earth.

Emmanuel B. Ehirim

Palmetto Publishing Group
Charleston, SC

I am Earth I am Your Home I am Fragile: Treat Me with Care
Copyright © 2019 by Emmanuel Ehirim

All rights reserved. No part of this book may be reproduced or transmitted in any form or by any means without written permission from the author.

Hardcover: 978-1-64111-583-4
Paperback: 978-1-64111-584-1
eBook: 978-1-64111-585-8

Printed in USA
Photos, images and pictures are from Pixabay free list of photos, images and pictures under the Pixabay license of use.

This book is written to give children, teenagers and adults too, an awareness of the problems and danger facing Earth and life on Earth.

It provides an insight of simple and practical things we can do to fix the problems created by humans and improve earth's land, water and atmosphere.

I am Earth, one of eight planets in the solar system. You live on me, planet Earth.

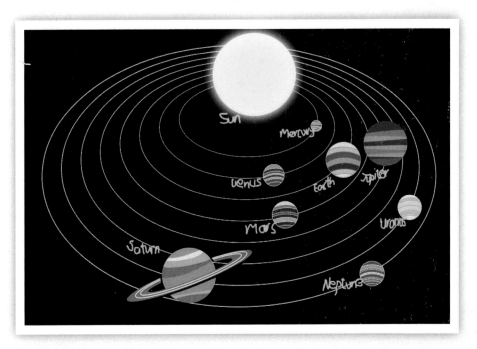

I am round like a ball and I am made up of land and water.

My land pieces are called continents and I have seven of them.

My continents are Asia, Africa, Europe, North America, South America, Australia and Antarctica.

My waters are called oceans and I have five of them. I also have other water sources: rivers, streams, creeks and lakes.

The oceans are Atlantic Ocean, Indian Ocean, Pacific Ocean, Southern Ocean and Arctic Ocean.

I have you, animals, fish, reptiles and insects living on me, and many plant life.

You are one of many, that is, billions of people living on me in all seven continents.

Plants, animals, birds, reptiles and insects all live on me. They are part of the land environment you live in.

Plants grow tall and small. Some plants bear fruits, vegetable and provide food for you.

Plants provide oxygen also which you, animals, birds and insects breathe to stay alive. Plants use and store the carbon dioxide you put in the atmosphere.

HEALTHY FOREST

FOREST HABITANTS

FOREST HABITANTS

Fish, ocean plants and other animals live in my waters.

These water sources are oceans, rivers, streams, creeks and lakes.

My waters in the northern area are frozen. These are ice sheets and are in the very cold part of me.

The water you drink comes from my rivers, streams, creeks and lakes.

OCEAN

STREAM

RIVER

RIVER BECOMING WATER FALL

I am Earth I am Your Home I am Fragile: Treat Me With Care

LAKE

CREEK

Animals, birds, insects and fish need everything you need to stay alive: air, water, food and a place to stay; habitat.

I want you to use everything I have: the land and water and live together with other living things: plants, animals, reptiles, fish and insects.

I am sad because I am not being treated nicely. I am getting sick and I will die if you do not treat me with care. When I die, you and all living thing, including plants, will die also.

My trees are being cut down excessively for fuel and timber to build houses.

DEFORESTED AREA FOR TIMBER AND FARMING

CUT DOWN TREES FOR TIMBERS

HOME BUILT FROM CUT DOWN TREE

My waters are being trashed with solid trash, plastics, dangerous chemicals from factories and you. My atmosphere, which is the air around me, is being polluted with poisonous gases and a lot of carbon dioxide too. The waste gas, which is carbon dioxide, comes from burning coal for electricity, trees for heating and cooking, gasoline to power automobile engines and small household equipment, like lawn mowers. These gases pollute my atmosphere and make it warmer.

You and others living on me are destroying me. I am getting sick and dying from air, water and land pollution.

DRAINAGE PIPE EMPTYING TRASH INTO OCEAN.

TRASH WASHED ASHORE

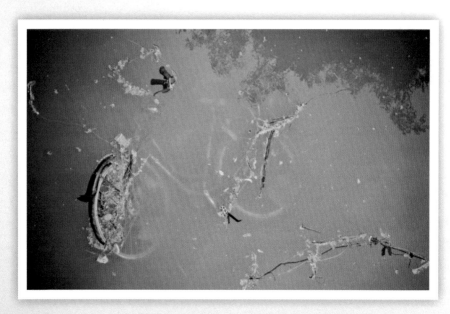

TRASH IN POLLUTED WATER

RIVER POLLUTED WITH CRUDE OIL SPILL

ANIMAL CAUGHT IN DUMPED TRASH

AIR POLLUTION FROM FACTORIES

AIR POLLUTION KILLING TREES

I have an area that is very cold; my north side has lots of snow on the mountains, on land and on the ocean. The snow on land and the ocean form ice sheets and glaciers.

The carbon dioxide from the vehicles and machines you use and the coal and wood you burn on land are getting into my atmosphere. This gas is also known as Green House gas. The gas is changing the climate in my atmosphere and causing my atmosphere to get warmer and therefore increasing the temperature on land. My waters are becoming warmer. My glaciers, ice sheets are melting at high rate because the atmosphere's temperature is rapidly increasing. The glaciers are breaking up into icebergs and melting, causing my oceans' levels to rise.

The rising oceans are causing land along the oceans to be covered by water. This is called flooding. Higher ocean levels threaten you. Coastal cities and local businesses, and a lot of plant and animal species are also threatened.

The land area you live on now will not be there in some years to come if the glaciers continue to melt and the ocean levels continue to increase.

The warming of my atmosphere that is earth's atmosphere is called Global Warming.

Global warming is causing some areas of me to get warmer and dryer. Some areas are becoming very hot and dry; you are not going to be able to live on them. Rivers, streams, creeks and lakes are drying up too because my atmosphere is getting hotter. The areas are becoming very dry and become desert land.

My rivers, creeks and lakes are drying up because of climate change. Desert land cannot support you, animals, reptiles, birds, insects and plant life. This means you will not be able to grow enough food to support yourself, and many others, including animals, reptiles, birds and insects.

**RIVER AND LAKE DRYING UP
BECAUSE OF WARMER CLIMATE**

You need to treat me nicely for you to continue to have me as your home. You have to preserve my lands, waters and atmosphere.

You have to preserve animals, birds, fish, reptiles and insects, like bees. Bees are, an important insect species helping to pollinate plant flowers, which become the fruits and food for you, animals, reptiles, birds and insects to eat and feed on.

GIRAFFE

WATER ANIMALS

BIRD

LARGE REPTILE

INSECT - ENDANGERED BEES

INSECT – DRAGON FLIES

I am Earth I am Your Home I am Fragile: Treat Me With Care

Do you know how to treat me nicely? I know you know how.

If you do not know, I will show you.

I am going to tell you about the things you need to do to keep me as your home for a long time to come.

Plant trees every time you cut down a tree and also when trees get old and die out. Plants absorb carbon dioxide in my atmosphere; they store it, use it to grow and produce oxygen in return. Oxygen is the gas you breathe to be alive.

Keep my water: oceans, rivers, streams, creeks and lakes clean. Stop putting carbon gases from vehicles, machines and, coal and wood burning into my atmosphere.

Do you know how to do that? I will show you.

This is what I need you to do: don't throw thrash on the ground or in drainages, in the oceans, streams, rivers, creeks and lakes.

Keep my waters: oceans, rivers, streams, creeks and lakes clean.

The trash you throw on the ground is washed into drainages, streams, rivers and creeks, and they end up in my oceans. Fish and other animals living in my oceans eat the plastic and solid trashes which go into them. These fish and other ocean living animals die after eating the plastic and solid trash.

Do not pour liquid waste, chemicals and unused medications into toilets, sinks, drainages, oceans, streams, rivers, creeks and lakes. The toxic chemicals sink into the soil and then pollute the soil and water inside of me.

Do you know what to do with your trash, liquid waste, chemicals and medications?

RECYCLE! RECYCLE! RECYCLE!

That is, collect your trash and, cans, plastic, paper and cardboards in special containers.

The trash and cans, plastics, papers and cardboards are collected and then reused to make new papers, cans, cardboard materials and plastic containers and wraps.

Dispose of unused medications properly. Take them to the nearest pharmacy where they can be disposed of safely.

Stop putting carbon gases from vehicles, machines and, coal and wood burning for energy into my atmosphere.

Use solar and wind energy for electric power to supply your home, businesses, offices and factories.

WIND TURBINES PRODUCING ELECTRIC ENERGY – CLEAN ENERGY

SOLAR CELLS PRODUCING ELECTRIC ENERGY – CLEAN ENERGY

Stop burning coal and diesel oil to generate electric energy.

Use automobiles and small equipments, e.g. lawn mowers and manufacturing factory equipments powered by electric engines.

ELECTRIC POWERED VEHICLES – CLEAN ENERGY

ELECTRIC POWERED MOTORIZED BIKE

I have been poisoned and contaminated by trash, plastic containers and bags, liquid waste, carbon dioxide (Co2), other toxic gases and chemical deposits for too long.

My trees are cut down constantly and not being replanted. My air and waters are being polluted everyday by carbon and other toxic gases from factories chimneys, machines and vehicles' exhaust pipes.

Make me feel better again and you will always be happy and healthy living on me; because you will have clean water to drink, clean and fresh air to breathe, nice parks with trees and clean rivers, streams, creeks, lakes and ocean to play in.

I AM YOUR HOME.

I can continue to exist for many, many years, if you help me stay healthy.

I AM FRAGILE:

Treat me with care.

WHEN I DIE, YOU WILL DIE TOO.

GLOSSARY

Atmosphere – The air around any location. An example of atmosphere is the ozone and other layers which make up the Earth's sky as we see it.

Carbon Dioxide (CO2) - A colorless, odorless gas that is produced through combustion and respiration.

Climate – The usual condition of the temperature, humidity, atmospheric pressure, wind, rainfall, and other meteorological elements in an area of the Earth's surface for a long time.

Deforestation – Is when forests are destroyed by cutting trees and not replanting them. It is sometimes what happens when people change lands into farms, ranches and settlements. Deforestation in the most common form, is obtaining wood for fuel and logs for construction. This destroys the habitat of many animals, leading to their death.

Global warning – Occurs when carbon dioxide (CO2) and other air pollutants and greenhouse gases collect in the atmosphere and absorb sunlight. It causes earth's climate change, which poses a serious threat to life on earth in the forms of widespread flooding and extreme weather.

Green house gas – A gas that absorbs infrared radiation (IR) and radiates heat in all directions. The primary greenhouse gases in Earth's atmosphere are water vapor, carbon dioxide, methane, nitrous oxide and ozone.

Ozone (O3) – A colorless, odorless reactive gas comprised of three oxygen atoms. It is found naturally in the earth's stratosphere, where it absorbs the ultraviolet component of incoming solar radiation that could be harmful to life on earth.

Solar power – Power obtained by harnessing the energy of the sun's rays.

Solar panel – A panel designed to absorb the sun's rays as a source of energy for generating electricity.

Wind power – Power obtained by harnessing the energy of the wind.

Wind turbine – A turbine having large vane wheel rotated by the wind to generate electricity.